Real Science-4-Kids

Kogs-4-Kids

Chemistry Connects to
Critical Thinking

Workbook Level I A

Rebecca W. Keller, Ph.D.

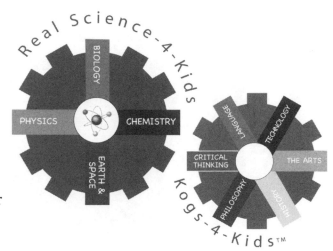

Cover design: David Keller
Opening page: David Keller
Illustrations: Janet Moneymaker, Rebecca Keller
Editing: Angie Sauberan
Page layout: Kimberly Keller

Real Science-4-Kids/ Kogs-4-Kids™ : Chemistry Connects to Critical Thinking: Level I A

ISBN: 9780979945939

Published by Gravitas Publications, Inc.
P.O. Box 4790
Albuquerque, NM
87196-4790
www.gravitaspublications.com

Printed in the United States of America

Special thanks to G.E. McEwan for valuable input.

Gravitas
Publications Inc.

I Introduction to Critical Thinking

I.1 What is critical thinking?

Humans think. You know this. Everyone, whether they seem to or not, thinks. You are thinking at this moment as you read these words. You might be thinking about where the person writing these words (me, the author) is going with this topic on thinking. You might be thinking that this a good way to start a paragraph on thinking—telling you that you think. You might be thinking that this is a lousy way to introduce a topic on thinking because it is obvious you are thinking, and you don't need anyone to tell you that. You might be thinking that you might

learn something new about thinking you never thought about. You might be thinking that you already know all there is to know about thinking, and you might be wondering why you are reading this in the first place. The point is that you are thinking right now. You think because you are human, and humans think.

But how do you think? Yes, there is a "biology of thinking" or a process that is going on in your brain as chemicals are being activated and deactivated as a result of your thinking. However, beyond biology, how do you think, and what do you think? What do you think exactly?

Do you think that you think clearly, or do you get lost in your thinking? Do you sometimes wonder if you are the only one thinking what you are thinking, or do you wonder if everyone thinks the way you think?

Can you think through a problem, or does it seem like thinking through a problem is the problem? Do you think that there are people who are just naturally good thinkers, like Albert Einstein? Do you think that these naturally good thinkers are the only thinkers who think and think, and with all their gifted thinking, discover amazing things? Or do you think that you could ever learn to think like Albert Einstein, and someday, think through and discover your own amazing things?

The fact is that almost anyone can learn to think like Albert Einstein. Yes, some people pick thinking up easily, but everyone can learn to think as well as Albert Einstein. Because everyone can learn to think well, everyone has within him or her new thoughts that could turn into new discoveries that are just as amazing as Albert Einstein's discoveries.

However, good thinking is hard work. Learning to think clearly and carefully takes training, patience, and practice. Thinking carefully with clarity, depth, precision, accuracy, and logic is *thinking critically*. Great scientists, like Albert Einstein, who discover amazing things about the world, have trained themselves to think critically. Critical thinking is the process of thinking in a certain way. Critical thinking is the process of thinking clearly, with accuracy and precision; of thinking carefully, with logic and depth; and of thinking open-mindedly, by examining points of view and acknowledging assumptions and biases within a given viewpoint. The point is that everyone can learn how to think critically if the time is taken to learn.

I.2 The tools of critical thinking

So what does it take to think critically? What are the nuts and bolts of critical thinking? Just like math or language or science, critical thinking has necessary tools and a method for using those tools.

There are two main activities we do all the time when we think. The first activity is gathering information or *collecting data*. As humans, our minds are constantly observing and collecting information about the world around us. We use our five senses to gather information about the world we live in. We are observing the height, size, weight, color, texture, and odor of the objects around us, and we are observing these qualities in relation to each other.

The second activity we do when we think is *drawing a conclusion* based on the information we've collected. We may conclude a building is too high to jump over, or an atom is too small to see with our eyes, or a boulder is too heavy to lift with our hands. However, what separates a critical thinker from a non-critical thinker is how she evaluates both the data she's collected and the conclusions she's drawn.

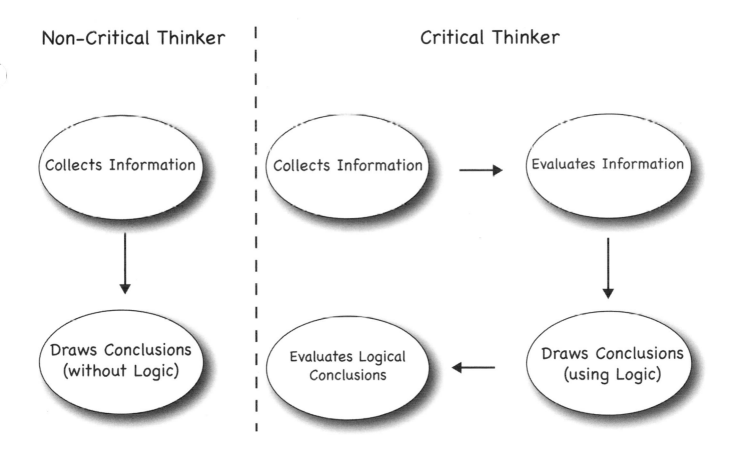

To evaluate both information and conclusions, the critical thinker must use the most important tools in the critical thinking toolbox: **questions**. To think critically, we must *ask questions* about the information or data we have collected. "Is it important?" "Is it relevant?" "Is it applicable?" "Is it significant?" But that's not enough. We must also *ask questions* about the conclusion we've drawn from the information we've collected. We need to ask the following types of questions: "Is the conclusion fair?" "Is is it logical?" "Is it reasonable?" and "Is it consistent with all the information collected?"

There are different kinds of questions (critical thinking tools) for different activities. There are tools for **Getting the Facts, Evaluating the Facts, Drawing a Conclusion,** and **Evaluating the Conclusion.**

I.3 Tools for getting the facts

When you first hear a statement or an argument, it is important to *get the facts*. If an officer has been called to an accident, the very first thing he does is get the facts. Who was involved? How were they

involved? Which car hit first? Which car hit second? Who was driving? Who wasn't driving? Exactly how fast was the first car going? When looking at something critically, it is important to collect as many facts as you can.

Tools for **Getting the Facts** include questions like ""Who?" "What?" "Where?" "When?" and "How?" The facts need to be **accurate**, **clear**, and **precise**. Questions that get to the **details** of facts, with words like "exactly," "how much," "what time," etc., help to clarify the facts.

I.4 Tools for evaluating the facts

Now that you know the facts, it is important to evaluate the facts. When an officer has collected all the facts for the accident, he needs to evaluate the facts. Evaluating facts is not as easy as it sounds because evaluations involve not only facts, but also involve opinions and preferences.

For example, one driver in the accident may claim that because there was a full moon, the accident was the other driver's fault. It might be a fact that there was a full moon, but is this fact relevant to the accident? Is it a significant fact concerning the accident? The officer has to evaluate the facts to find out if they are facts that should or should not be used to draw a conclusion.

Tools for **Evaluating the Facts** include questions that explore the **relevance** and **significance** of the facts and questions that explore whether or not the facts are **substantial**, **crucial**, or **applicable** to the conclusion.

I.5 Tools for drawing a conclusion-using logic

Now that we have collected the facts and evaluated the facts, we can "draw a conclusion." A conclusion is a statement that sums up all of the information collected in order to make a point or a decision. But how do you know if the conclusion

you've made is valid and consistent, or logically flawed? For example, one driver might not like men in flowered shirts. This driver might want to say that it was not the moon that caused the accident, but that it was the man in the flowered shirt that caused the accident because "men in flowered shirts always cause accidents." Is this true, or is the driver making a logical error?

Tools for **Drawing a Conclusion** use logic (a method that investigates arguments) to help the critical thinker avoid making errors by exploring **validity**, **consistency**, and **logical flaws**.

I.6 Tools for evaluating a conclusion

Sometimes it's not enough to have a logical conclusion. Sometimes it is necessary to evaluate your conclusion. We need to ask the following types of questions: "Is my conclusion fair?" "Has my conclusion taken into

account all the information available?" "Is my conclusion reasonable?" and "Is there more information that should be considered?" For example, the officer may conclude that the moon did not cause the accident, and that the man in the flowered shirt did not cause the accident, but that instead, neither man was watching where he was going. One was looking at the moon, and the other was fixing a button on his shirt; so they are both at fault. But does that conclusion take into account all the information available, or is there more information that must be considered before the officer can make a fair conclusion?

Tools for **Evaluating a Conclusion** include questions that explore the **fairness, reasonableness, depth,** and **breadth** of a conclusion.

I.7 Putting it all together- critical thinking

In summary, the four main types of critical thinking tools are: **Getting the Facts, Evaluating the Facts, Drawing a Conclusion using Logic,** and **Evaluating a Conclusion.**

As we've mentioned, asking questions is the key for critical thinking, and it is important to ask questions that incorporate all of the critical thinking tools we've discussed. It is important that we ask questions not just of other people's thinking, but that we also challenge, and ask questions of, our own thinking.

The critical thinking tools we've discussed are different kinds of questions that explore different aspects of the information gathered, and that explore different aspects of the conclusions drawn from that information. Throughout this workbook, we will be asking questions using all of the critical thinking tools.

Finally, one of the most important questions you can ask another person is, "Let me understand what you are saying. Are you saying...?" Then in different words, repeat what you think the other person is saying, or repeat what you think you are saying in a different way. To admit you may not understand what someone else is saying is a way to open up more critical thinking questions.

I.8 Building a critical thinking lens

We have been talking about "critical thinking tools," but what exactly do all of these critical thinking tools look like together? One way to

envision all of the critical thinking tools is to think about a lens. If our eyes do not function properly, a lens helps us see objects more clearly. In the same way, a **critical thinking lens** can help you think through problems more clearly.

Constructing a critical thinking lens is not very

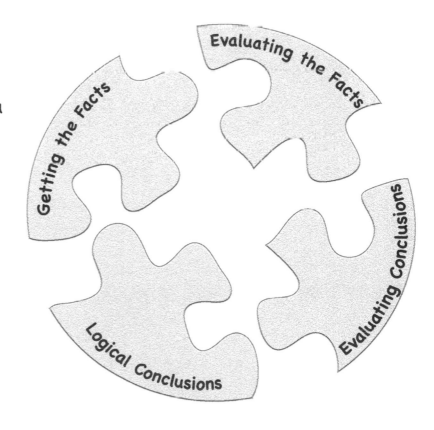

difficult. It amounts to asking questions using the four critical thinking tools we have been learning. As you improve your ability to ask good questions, your critical thinking lens will improve. A critical thinking lens can help you decide what kinds of statements are scientifically valid, and what kinds of statements may not be scientifically valid.

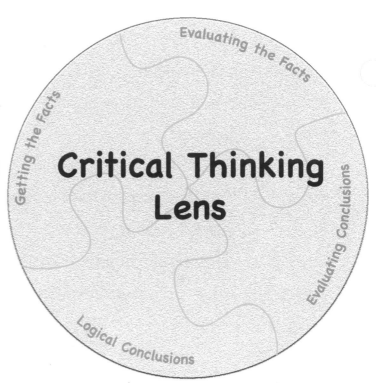

I.9 Summary

❶ Critical thinking tools are **questions**.

❷ There are four main types of critical thinking tools (questions): **Getting the Facts, Evaluating the Facts, Drawing a Conclusion using Logic,** and **Evaluating a Conclusion.**

❸ Tools for **Getting the Facts** include questions like "Who?" "What?" "Where?" "When?" and "How?"

❹ Tools for **Evaluating the Facts** include the following types of questions: "Is this fact relevant or significant?" "Is this fact substantial, crucial, and applicable?" and "Does it support the conclusion?"

❺ Tools for **Drawing a Conclusion** use logic to help the critical thinker to avoid making errors by asking: "Is this valid and consistent with other information?" and "Are there any logical flaws in this conclusion?"

❻ Tools for **Evaluating a Conclusion** include the following types of questions: "Is this fair and reasonable?" and "Does my conclusion have the necessary depth and breadth?"

I.10 Discussion questions

Look at the following scientific claim:

The moon is made of green cheese.

Look at the critical thinking lens on page 15.

1. Can you pick out two **Getting the Facts** questions?

❶ _W_____

_____

❷ _____

_____

2. Can you pick out two **Evaluating the Facts** questions?

❶ _W_____

_____

❷ _____

_____

3. Based on the critical thinking lens, do you think that the moon is
   made of green cheese? Why or why not?

_____

_____

_____

_____

_____

_____

_____

_____

_____

_____

_____

_____

_____

_____

_____

_____

4. Have you considered enough information to draw that conclusion? (Does your answer have depth and breadth?). If not, what other information should you consider?

_____

_____

_____

_____

_____

_____

_____

_____

_____

_____

_____

_____

_____

_____

_____

_____

_____

_____

_____

_____

_____

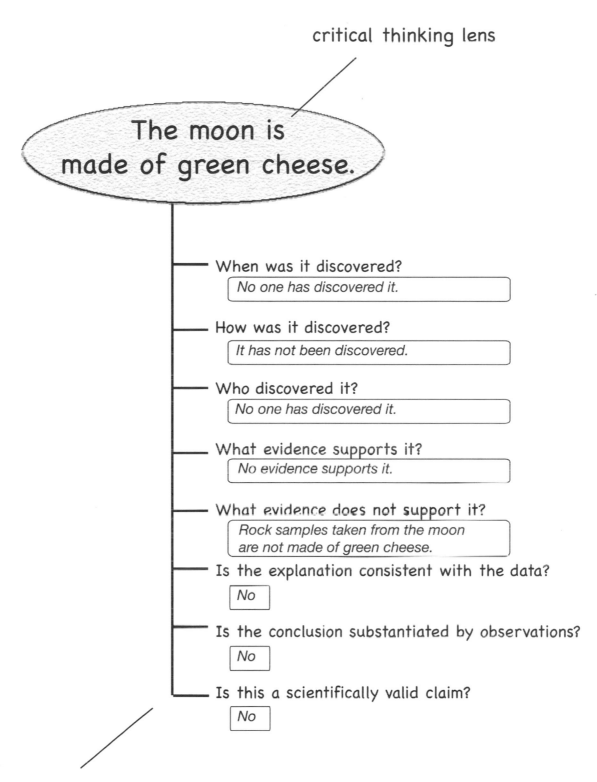

critical thinking lens

The moon is made of green cheese.

When was it discovered?
No one has discovered it.

How was it discovered?
It has not been discovered.

Who discovered it?
No one has discovered it.

What evidence supports it?
No evidence supports it.

What evidence does not support it?
Rock samples taken from the moon are not made of green cheese.

Is the explanation consistent with the data?
No

Is the conclusion substantiated by observations?
No

Is this a scientifically valid claim?
No

These are the thinking tools (the questions) that, together, make the critical thinking lens.

1

The Atom
Critical Thinking

1.1 The atom

In chapter 1 of Chemistry Level I, you learned all about the **atom**. You also read about the history of the **atom** and how **atoms** were discovered. Now you will construct a critical thinking lens to evaluate this scientific claim:

Matter is made of atoms.

[Note: You will need to consider the information you find in your textbook and other resources as "facts" in order to complete this exercise. However, know that the "facts" in your textbook and in other resources are really conclusions that are based on many other facts that have been collected over years of investigation and that have been evaluated by many scientists. As a critical thinker, you are encouraged to examine all "facts" and to evaluate them for yourself, even those facts that have already been evaluated by other scientists.]

1.2 Gathering the tools

First, we need to gather the critical thinking tools. The four types of tools we will be using are as follows:

Tools for Getting the Facts
> *questions regarding clarity, precision, accuracy, and detail*

Tools for Evaluating the Facts
> *questions regarding significance, relevancy, and application*

Tools for Drawing Conclusions (using logic)
> *questions regarding logical validity, consistency, and flaws*

Tools for Evaluating Conclusions
> *questions regarding fairness, depth, breadth, and reasonableness*

A. Tools for Getting the Facts

questions regarding clarity, precision, accuracy, and detail

Answer the following **Getting the Facts** questions for the statement:

Matter is made of atoms.

1. Q: Who discovered the atom, and in what year was it discovered?

   A: _____

   _____

   _____

2. Q: What are atoms made of?

   A: _____

   _____

   _____

B. Tools for Evaluating the Facts

questions regarding significance, relevancy, and application

Answer the following **Evaluating the Facts** questions for the statement: *Matter is made of atoms.*

1. Q: Is the fact that the atom was discovered significant to the argument that *matter is made of atoms*?

   A: _____

   _____

   _____

2. Q: Is the data that says what atoms are made of relevant to the argument that *matter is made of atoms*?

   A: _____

   _____

   _____

C. Tools for Drawing a Conclusion (using Logic)

questions regarding logical validity, consistency, and flaws

In this section, you will learn how to recognize valid arguments and logical fallacies. A logical fallacy is an inaccurate way to formulate an argument. In this chapter, you will be introduced to the logical fallacy called **equivocation**.

Logical Fallacy: Equivocation (fallacy of ambiguity)

Definition: A word or phrase used in the argument that is not clearly defined, or that changes definition during the argument.

Example: Because metal sinks in water, you can't make a sink from metal.

To prevent committing the logical fallacy of *equivocation*, the definitions of all of the terms in the argument must stay the same. We are using statements of fact to support this argument:

Matter is made of atoms.

Write a definition for **matter.**_____

_____

_____

Write a definition for **atom.** _____

_____

_____

Look at the following two statements and the *conclusion*.

    1. Atoms contain protons, neutrons, and electrons.

    2. Matter contains protons, neutrons, and electrons.

Therefore (*conclusion*),

    Matter is made of atoms.

Determine if the definitions for **matter** and **atom** stay the same.

Q: Does the conclusion that *matter is made of atoms* commit the fallacy of equivocation based on the information you have?

A: ☐ Yes   ☐ No

D. Tools for Evaluating the Conclusion

    questions regarding fairness, depth, breadth, and reasonableness

Answer the following **Evaluating the Conclusion** questions for the statement: *Matter is made of atoms.*

1. Q: Is the fact that the atom was discovered significant to the argument that *matter is made of atoms*?

   A:

_____

_____

_____

2. Q: Is the data that says what atoms are made of relevant to the argument that *matter is made of atoms*?

   A:

_____

_____

_____

1.3 Building the critical thinking lens

You have gathered the facts, evaluated the facts, checked the conclusion using logic, and evaluated the conclusion:

Matter is made of atoms.

Next, put all of the facts, evaluations, and logical checks together to construct a critical thinking lens.

Write the statement you are evaluating in the critical thinking lens.

Matter is made of atoms.

Write the two **Getting the Facts** critical thinking questions.

❶

❷

Write the two **Evaluating the Facts** questions.

❶

❷

Write two **Drawing a Conclusion using Logic** statements that don't commit a logical fallacy.

❶

❷

Write the two **Evaluating the Conclusion** questions.

❶

❷

1.4 Using the critical thinking lens

Look at the critical thinking lens you constructed and think about the answers to the critical thinking questions in your critical thinking lens. Do you think that the statement *"matter is made of atoms"* is a good scientific argument?

☐ Yes   ☐ No

Why or why not?

_____

_____

_____

_____

_____

_____

_____

_____

_____

_____

_____

_____

_____

_____

_____

1.5 Now you try

You run into a scientist on the street, and you start talking. He tells you his scientific opinion:

The cow jumped over the moon.

Evaluate his argument by constructing a critical thinking lens.

Tools for Getting the Facts

Write two questions regarding clarity, precision, accuracy, and detail.

❶ _____

_____

❷ _____

_____

Tools for Evaluating the Facts

Write two questions regarding significance, relevancy, and application.

❶ _____

_____

❷ _____

_____

Tools for Drawing Conclusions (using logic)

Write two questions regarding logical validity, consistency, and flaws.

❶ _____

_____

❷ _____

_____

Tools for Evaluating Conclusions

Write two questions regarding fairness, depth, breadth, and reasonableness.

❶ _____

_____

❷ Cows don't _____

1.6 Make your own

Using the questions you came up with in section 1.5, construct your own critical thinking lens.

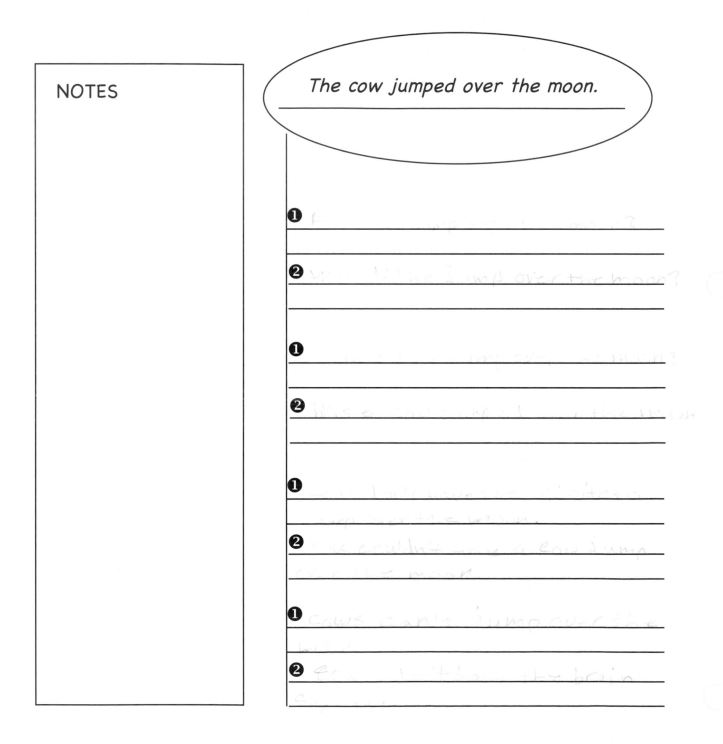

NOTES

The cow jumped over the moon.

❶

❷

❶

❷

❶

❷

❶

❷

Molecules
Critical Thinking

2.1 Molecules

In chapter 2 of Chemistry Level I, you learned about **molecules**. You also read about the history of **molecules** and how **molecules** were discovered. Now you will construct a critical thinking lens to evaluate this scientific claim:

Hydrogen forms only one bond.

[Note: You will need to consider the information you find in your textbook and other resources as "facts" in order to complete this exercise. However, know that the "facts" in your textbook and in other resources are really conclusions that are based on many other facts that have been collected over years of investigation and that have been evaluated by many scientists. As a critical thinker, you are encouraged to examine all "facts" and to evaluate them for yourself, even those facts that have already been evaluated by other scientists.]

2.2 Gathering the tools

First, we need to gather the critical thinking tools. The four types of tools we will be using are as follows:

Tools for Getting the Facts

questions regarding clarity, precision, accuracy, and detail

Tools for Evaluating the Facts

questions regarding significance, relevancy, and application

Tools for Drawing Conclusions (using logic)

questions regarding logical validity, consistency, and flaws

Tools for Evaluating Conclusions

questions regarding fairness, depth, breadth, and reasonableness

A. Tools for Getting the Facts

questions regarding clarity, precision, accuracy, and detail

Answer the following **Getting the Facts** questions for the statement:

Hydrogen forms only one bond.

1. Q: What makes a bond?

    A:

_____

_____

_____

2. Q: How many electrons does hydrogen have?

    A:

_____

_____

_____

B. Tools for Evaluating the Facts

questions regarding significance, relevancy, and application

Answer the following **Evaluating the Facts** questions for the

statement: *Hydrogen forms only one bond.*

1. Q: Is the fact that hydrogen has one electron significant to the
      argument that *hydrogen forms only one bond?*

    A:

_____

_____

2. Q: Is the data that says how bonds are formed relevant to the
      argument that *hydrogen forms only one bond?*

    A:

_____

_____

_____

C. Tools for Drawing a Conclusion (using Logic)

questions regarding logical validity, consistency, and flaws

In this section, you will learn how to recognize valid arguments and logical fallacies. A logical fallacy is an inaccurate way to formulate an argument. In this chapter, you will be introduced to the logical fallacy called **appeal to force**.

Logical Fallacy: Appeal to Force (*Argumentum Ad Baculum*)

Definition: "Might-makes-right"- The argument uses force or the threat of force to make the listener accept the conclusion.

Example: The bike belongs to me because Mom will find out that you broke her vase if you don't agree with me.

To prevent committing the logical fallacy of *appeal to force*, force or the threat of force cannot be used as an argument for this statement:

Hydrogen forms only one bond.

Evaluate the following arguments. Mark those using an *appeal to force*.

☐ Water contains hydrogen that is bonded to oxygen only once.

☐ Everyone in class who does not agree that hydrogen forms only one bond will not get a passing grade.

☐ Only people who agree that hydrogen forms only one bond will get water to drink.

☐ Hydrogen forms one bond in the following molecules: HCN, HCl, H_2SO_4.

☐ Hydrogen forms only one bond. Trust me, or I won't speak to you again.

☐ Hydrogen has only one electron, so it can form only one bond.

D. Tools for Evaluating the Conclusion

questions regarding fairness, depth, breadth, and reasonableness

Answer the following **Evaluating the Conclusion** questions for the statement: *Hydrogen forms only one bond.*

1. Q: Is the fact that hydrogen has only one electron significant to the argument that *hydrogen forms only one bond*?

   A: _____

   _____

   _____

2. Q: Is it reasonable to conclude that because bonds are formed with electrons, *hydrogen forms only one bond* because it has only one electron?

   A: _____

   _____

   _____

2.3 Building the critical thinking lens

You have gathered the facts, evaluated the facts, checked the conclusion using logic, and evaluated the conclusion:

Hydrogen forms only one bond.

Next put all of the facts, evaluations, and logical checks together to construct a critical thinking lens.

Write the statement you are evaluating in the critical thinking lens.

Hydrogen forms only one bond.

Write the two **Getting the Facts** critical thinking questions.

❶

❷

Write the two **Evaluating the Facts** questions.

❶

❷

Write two **Drawing a Conclusion using Logic** statements that don't commit a logical fallacy.

❶

❷

Write the two **Evaluating the Conclusion** questions.

❶

❷

2.4 Using the critical thinking lens

Look at the critical thinking lens you constructed and think about the answers to the critical thinking questions in your critical thinking lens. Do you think that the statement "*hydrogen forms only one bond*" is a good scientific argument?

Yes ☐   No ☐

Why or why not?

_____

_____

_____

_____

_____

_____

_____

_____

_____

_____

_____

_____

_____

2.5 Now you try

You run into a scientist on the street, and you start talking. He tells you his scientific opinion:

The sky is blue.

Evaluate his argument by constructing a critical thinking lens.

Tools for Getting the Facts

Write two questions regarding clarity, precision, accuracy, and detail.

❶ _____

_____

❷ _____

_____

Tools for Evaluating the Facts

Write two questions regarding significance, relevancy, and application.

❶ _____

_____

❷ _____

_____

Tools for Drawing Conclusions (using logic)

Write two questions regarding logical validity, consistency, and flaws.

❶ _____

_____

❷ _____

_____

Tools for Evaluating Conclusions

Write two questions regarding fairness, depth, breadth, and reasonableness.

❶ _____

_____

❷ _____

_____

2.6 Make your own

Using the questions you came up with in section 2.5, construct your own critical thinking lens.

NOTES

The sky is blue.
_____

❶ _____

❷ _____

❶ _____

❷ _____

❶ _____

❷ _____

❶ _____

❷ _____

3 Chemical Reactions
Critical Thinking

3.1 Chemical reactions

3.2 Gathering the tools

3.3 Building the critical thinking lens

3.4 Using the critical thinking lens

3.5 Now you try

3.6 Make your own

3.1 Chemical reactions

In chapter 3 of Chemistry Level I, you learned about **chemical reactions**. You also read about the history of **chemical reactions** and how **chemical reactions** were discovered. Now you will construct a critical thinking lens to evaluate the following scientific claim:

Sodium and chlorine combine to form table salt.

[Note: You will need to consider the information you find in your textbook and other resources as "facts" in order to complete this exercise. However, know that the "facts" in your textbook and in other resources are really conclusions that are based on many other facts that have been collected over years of investigation and that have been evaluated by many scientists. As a critical thinker, you are encouraged to examine all "facts" and to evaluate them for yourself, even those facts that have already been evaluated by other scientists.]

3.2 Gathering the tools

First, we need to gather the critical thinking tools. The four types of tools we will be using are as follows:

Tools for Getting the Facts
questions regarding clarity, precision, accuracy, and detail

Tools for Evaluating the Facts
questions regarding significance, relevancy, and application

Tools for Drawing Conclusions (using logic)
questions regarding logical validity, consistency, and flaws

Tools for Evaluating Conclusions
questions regarding fairness, depth, breadth, and reasonableness

A. Tools for Getting the Facts

questions regarding clarity, precision, accuracy, and detail

Answer the following **Getting the Facts** questions for the statement:

Sodium and chlorine combine to form table salt.

1. Q: What atoms are found in normal table salt?

   A: _____

   _____

   _____

2. Q: What kind of reaction between sodium and chlorine form table salt?

   A: _____

   _____

   _____

B. Tools for Evaluating the Facts

questions regarding significance, relevancy, and application

Answer the following **Evaluating the Facts** questions for the

statement: *Sodium and chlorine combine to form table salt.*

1. Q: Is the fact that chlorine and sodium are in table salt
   significant to the argument that *sodium and chlorine combine
   to form table salt?*

   A: _____

   _____

   _____

2. Q: Is the fact that a combination reaction occurs between sodium
   and chlorine applicapable to the argument, *sodium and chlorine
   combine to form table salt?*

   A: _____

   _____

   _____

C. Tools for Drawing a Conclusion (using Logic)

questions regarding logical validity, consistency, and flaws

In this section, you will learn how to recognize valid arguments and logical fallacies. A logical fallacy is an inaccurate way to formulate an argument. In this chapter, you will be introduced to the logical fallacy called **personal attack**.

Logical Fallacy: Personal Attack (*Ad Hominem*)

Definition: A person makes abusive remarks about the person making the argument as evidence against the argument the person is making.

Example: All boys who wear yellow don't know anything. Because Bobby wears yellow, what he says is not true.

To prevent committing the logical fallacy of *personal attack,* it is best to stick with the content of the argument and not focus on the person who is making the argument. Evaluate the following arguments for this statement:

Sodium and chlorine combine to form table salt.

Mark those using a *personal attack.*

☐ Sodium chloride is a mineral and melts at 803°C.

☐ People who test table salt wear too much pink. You can't trust people in pink. Therefore, table salt does not have sodium and chlorine in it.

☐ Sodium and chlorine are known to react with each other in a combination reaction.

☐ Sir Humphrey Davy discovered sodium, but boys named Humphrey can't do science, so sodium does not exist. Therefore, table salt does not contain sodium.

☐ Chlorine was discovered by a Swedish chemist, but because Swedish chemists wear too much blue and yellow, he can't be right.

D. Tools for Evaluating the Conclusion

questions regarding fairness, depth, breadth, and reasonableness

Answer the following **Evaluating the Conclusion** questions for the statement: *Sodium and chlorine combine to form table salt.*

1. Q: Is it fair to say that because there exists a compound, sodium chloride, that is used to flavor food, *sodium and chlorine combine to form table salt?*

   A: _____

   _____

   _____

2. Q: Is it reasonable to conclude that because people in pink don't like salt and can't be trusted, *sodium and chlorine combine to form table salt?*

   A: _____

   _____

   _____

3.3 Building the critical thinking lens

You have gathered the facts, evaluated the facts, checked the conclusion using logic, and evaluated the conclusion:

Sodium and chlorine combine to form table salt.

Next, put all of the facts, evaluations, and logical checks together to construct a critical thinking lens.

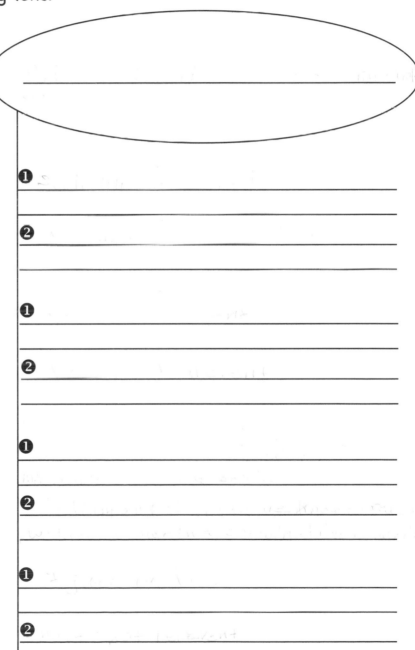

Write the statement you are evaluating in the critical thinking lens.

Write the two **Getting the Facts** critical thinking questions.

❶

❷

Write the two **Evaluating the Facts** questions.

❶

❷

Write two **Drawing a Conclusion using Logic** statements that do not commit a logical fallacy.

❶

❷

Write the two **Evaluating the Conclusion** questions.

❶

❷

3.4 Using the critical thinking lens

Look at the critical thinking lens you constructed and think about the answers to the critical thinking questions in your critical thinking lens. Do you think that the statement "*sodium and chlorine combine to form table salt*" is a good scientific argument?

Yes ☐   No ☐

Why or why not?

_____

_____

_____

_____

_____

_____

_____

_____

_____

_____

_____

_____

_____

_____

3.5 Now you try

You run into a scientist on the street, and you start talking. He tells you his scientific opinion:

Table salt dissolves in water.

Evaluate his argument by constructing a critical thinking lens.

Tools for Getting the Facts

Write two questions regarding clarity, precision, accuracy, and detail.

❶ _____

_____

❷ _____

_____

Tools for Evaluating the Facts

Write two questions regarding significance, relevancy, and application.

❶ _____

_____

❷ _____

_____

Tools for Drawing Conclusions (using logic)

Write two questions regarding logical validity, consistency, and flaws.

❶ _____

_____

❷ _____

_____

Tools for Evaluating Conclusions

Write two questions regarding fairness, depth, breadth, and reasonableness.

❶ _____

_____

❷ _____

_____

3.6 Make your own

Using the questions you came up with in section 3.5, construct your own critical thinking lens.

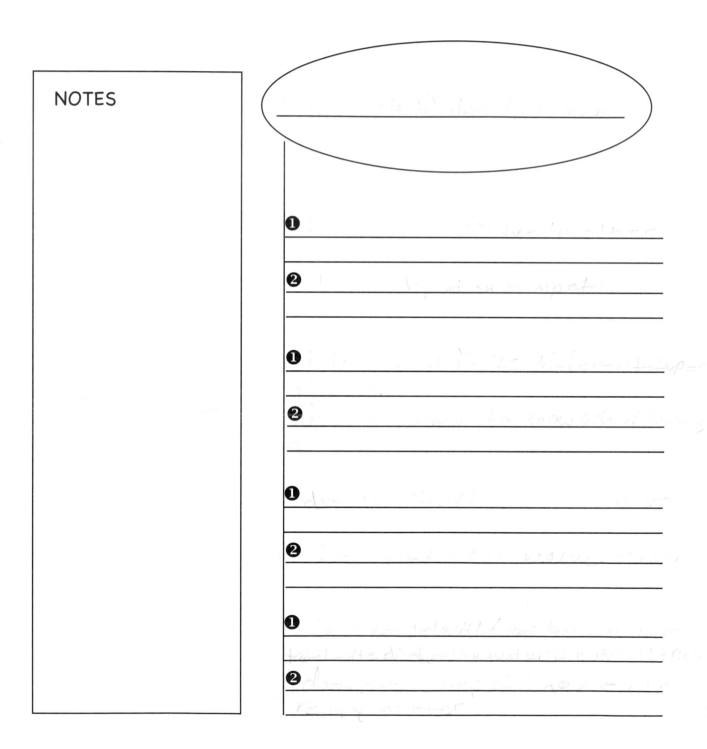

NOTES

❶

❷

❶

❷

❶

❷

❶

❷

4 Acids and Bases
Critical Thinking

4.1 Acids and bases

In chapter 4 of Chemistry Level I, you learned about **acids and bases**. You also read about the history of **acids and bases** and how **acids and bases** were discovered. Now you will construct a critical thinking lens to evaluate the following scientific claim:

Vinegar is an acid.

[Note: You will need to consider the information you find in your textbook and other resources as "facts" in order to complete this exercise. However, know that the "facts" in your textbook and in other resources are really conclusions that are based on many other facts that have been collected over years of investigation and that have been evaluated by many scientists. As a critical thinker, you are encouraged to examine all "facts" and to evaluate them for yourself, even those facts that have already been evaluated by other scientists.]

4.2 Gathering the tools

First, we need to gather the critical thinking tools. The four types of tools we will be using are as follows:

Tools for Getting the Facts
> *questions regarding clarity, precision, accuracy, and detail*

Tools for Evaluating the Facts
> *questions regarding significance, relevancy, and application*

Tools for Drawing Conclusions (using logic)
> *questions regarding logical validity, consistency, and flaws*

Tools for Evaluating Conclusions
> *questions regarding fairness, depth, breadth, and reasonableness*

A. Tools for Getting the Facts

questions regarding clarity, precision, accuracy, and detail

Answer the following **Getting the Facts** questions for the statement:

Vinegar is an acid.

1. Q: What atoms are found in vinegar?

   A:
   _____

   _____

   _____

2. Q: What is the pH of vinegar?

   A:
   _____

   _____

   _____

B. Tools for Evaluating the Facts

questions regarding significance, relevancy, and application

Answer the following **Evaluating the Facts** questions for the statement: *Vinegar is an acid.*

1. Q: Is the fact that vinegar contains a hydrogen atom attached to an oxygen atom significant to the argument that *vinegar is an acid*?

   A:
   _____

   _____

   _____

2. Q: Is the fact that the pH of vinegar is below 7 relevant to the argument that *vinegar is an acid*?

   A:
   _____

   _____

C. Tools for Drawing a Conclusion (using Logic)

questions regarding logical validity, consistency, and flaws

In this section, you will learn how to recognize valid arguments and logical fallacies. A logical fallacy is an inaccurate way to formulate an argument. In this chapter, you will be introduced to the logical fallacy called **straw man**.

Logical Fallacy: Straw Man

Definition: The straw man fallacy is committed when a person misrepresents or exaggerates the other person's position.

Example: History is only about the past. Because my teacher said we all live in the present, my teacher is wrong to give me a C on a test about the past.

To prevent committing the logical fallacy of *straw man,* make sure that the argument has not been misrepresented. Evaluate the following arguments for this statement:

Vinegar is an acid.

Mark those using a *straw man.*

☐ Acids burn people. Because vinegar does not burn, it is not an acid.

☐ Vinegar contains a hydrogen that can come off in water (hydrolyze), and is therefore, an acid in water.

☐ No foods are acids. Because vinegar is a food, it is not an acid.

☐ Vinegar can be used to clean a coffee maker. Because all cleaners are soaps, vinegar is a soap, not an acid.

☐ The pH of vinegar in water is below 7. Therefore, vinegar is an acid.

D. Tools for Evaluating the Conclusion

questions regarding fairness, depth, breadth, and reasonableness

Answer the following **Evaluating the Conclusion** questions for the statement: *Vinegar is an acid.*

1. Q: Is it reasonable to say that because the pH of vinegar is below 7, *vinegar is an acid?*

    A:

    _____

    _____

    _____

2. Q: Is it fair to conclude that because acids burn, and vinegar doesn't burn, it is not true that *vinegar is an acid?*

    A:

    _____

    _____

    _____

4.3 Building the critical thinking lens

You have gathered the facts, evaluated the facts, checked the conclusion using logic, and evaluated the conclusion:

Vinegar is an acid.

Next, put all of the facts, evaluations, and logical checks together to construct a critical thinking lens.

Write the statement you are evaluating in the critical thinking lens.

Write the two **Getting the Facts** critical thinking questions.

❶ _____

❷ _____

Write the two **Evaluating the Facts** questions.

❶ _____

❷ _____

Write two **Drawing a Conclusion using Logic** statements that don't commit a logical fallacy.

❶ _____

❷ _____

Write the two **Evaluating the Conclusion** questions.

❶ _____

❷ _____

4.4 Using the critical thinking lens

Look at the critical thinking lens you constructed and think about the answers to the critical thinking questions in your critical thinking lens. Do you think that the statement "*vinegar is an acid*" is a good scientific argument?

☐ Yes   ☐ No

Why or why not?

_____

_____

_____

_____

_____

_____

_____

_____

_____

_____

_____

_____

_____

_____

4.5 Now you try

You run into a scientist on the street, and you start talking. He tells you his scientific opinion:

Apples are green.

Evaluate his argument by constructing a critical thinking lens.

Tools for Getting the Facts

Write two questions regarding clarity, precision, accuracy, and detail.

❶ _____

_____

❷ _____

_____

Tools for Evaluating the Facts

Write two questions regarding significance, relevancy, and application.

❶ _____

_____

❷ _____

_____

Tools for Drawing Conclusions (using logic)

Write two questions regarding logical validity, consistency, and flaws.

❶ _____

_____

❷ _____

_____

Tools for Evaluating Conclusions

Write two questions regarding fairness, depth, breadth, and reasonableness.

❶ _____

❷ _____

_____

4.6 Make your own

Using the questions you came up with in section 4.5, construct your own critical thinking lens.

NOTES

5 Acids–Base Reactions
Critical Thinking

5.1 Acid-base reactions

5.2 Gathering the tools

5.3 Building the critical thinking lens

5.4 Using the critical thinking lens

5.5 Now you try

5.6 Make your own

5.1 Acid-base reactions

In chapter 5 of Chemistry Level I, you learned about **acid-base reactions**. You also read about the history of **acid-base reactions** and how **acid-base reactions** were discovered. Now you will construct a critical thinking lens to evaluate the following scientific claim:

Antacids are bases that neutralize stomach acid.

[Note: You will need to consider the information you find in your textbook and other resources as "facts" in order to complete this exercise. However, know that the "facts" in your textbook and in other resources are really conclusions that are based on many other facts that have been collected over years of investigation and that have been evaluated by many scientists. As a critical thinker, you are encouraged to examine all "facts" and to evaluate them for yourself, even those facts that have already been evaluated by other scientists.]

5.2 Gathering the tools

First, we need to gather the critical thinking tools. The four types of tools we will be using are as follows:

Tools for Getting the Facts
   questions regarding clarity, precision, accuracy, and detail
Tools for Evaluating the Facts
   questions regarding significance, relevancy, and application
Tools for Drawing Conclusions (using logic)
   questions regarding logical validity, consistency, and flaws.
Tools for Evaluating Conclusions
   questions regarding fairness, depth, breadth, and reasonableness

A. Tools for Getting the Facts

questions regarding clarity, precision, accuracy, and detail

Answer the following **Getting the Facts** questions for the statement:

Antacids are bases that neutralize stomach acid.

1. Q: What is in an antacid?

    A: _____

    _____

    _____

2. Q: When you get a stomach ache, what might a doctor tell you?

    A: _____

    _____

    _____

B. Tools for Evaluating the Facts

questions regarding significance, relevancy, and application

Answer the following **Evaluating the Facts** questions for the
statement: *Antacids are bases that neutralize stomach acid*

1. Q: Is the fact that antacids contain calcium carbonate or
       aluminum hydroxide significant to the argument that *antacids
       are bases that neutralize stomach acid*?

    A: _____

    _____

    _____

2. Q: Is the fact that a doctor might tell you to take an antacid for a
       stomach ache relevant to the argument that *antacids are bases
       that neutralize stomach acid*?

    A: _____

    _____

    _____

C. Tools for Drawing a Conclusion (using Logic)

questions regarding logical validity, consistency, and flaws

In this section, you will learn how to recognize valid arguments and logical fallacies. A logical fallacy is an inaccurate way to formulate an argument. In this chapter, you will be introduced to the logical fallacy called **appeal to popularity**.

Logical Fallacy: Appeal to Popularity (*Ad Populum*)

Definition: The appeal to popularity fallacy is committed when an argument is accepted as true simply because many people believe it is true.

Example: Because most kids think that eating brussel sprouts will make them sick, brussel sprouts must be unhealthy.

To prevent committing the logical fallacy of *appeal to popularity,* make sure that the argument is not being accepted because it is popular. Evaluate the following arguments for this statement:

Antacids are bases that neutralize stomach acid.

Mark those using an *appeal to popularity.*

☐ Antacids contain carbonate or hydroxide, which are bases.

☐ Everybody uses antacids to neutralize stomach acid, so they must be basic.

☐ All of the smartest people believe antacids are bases that neutralize stomach acid. Therefore, antacids must neutralize acids.

☐ Some antacids contain calcium carbonate or magnesium hydroxide, which neutralize acid.

☐ Doctors recommend antacids for stomach acid. Because doctors recommend antacids, antacids must be basic.

D. Tools for Evaluating the Conclusion

questions regarding fairness, depth, breadth, and reasonableness

Answer the following **Evaluating the Conclusion** questions for the statement:

Antacids are bases that neutralize stomach acid.

1. Q: Is it fair to say that because antacids contain carbonate, *antacids are bases that neutralize stomach acid*?

    A:
    _____
    _____
    _____

2. Q: Do you think it is reasonable to conclude that because doctors recommend antacids for stomach aches, *antacids are bases that neutralize stomach acid*?

    A:
    _____
    _____
    _____

5.3 Building the critical thinking lens

You have gathered the facts, evaluated the facts, checked the conclusion using logic, and evaluated the conclusion:

Antacids are bases that neutralize stomach acid.

Next, put all of the facts, evaluations, and logical checks together to construct a critical thinking lens.

Write the statement you are evaluating in the critical thinking lens.

Write the two **Getting the Facts** critical thinking questions.

Write the two **Evaluating the Facts** questions.

Write two **Drawing a Conclusion using Logic** statements that don't commit a logical fallacy.

Write the two **Evaluating the Conclusion** questions.

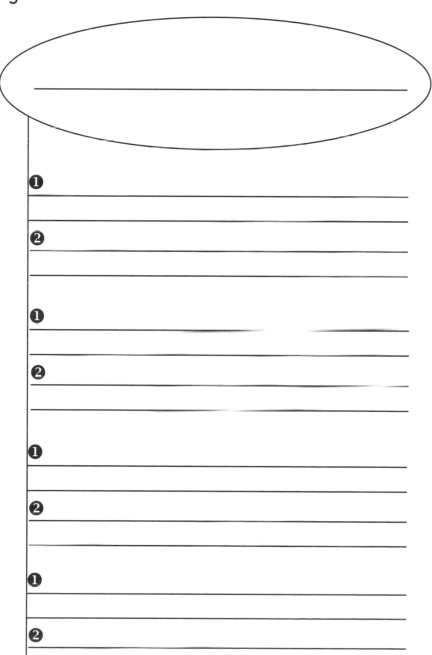

5.4 Using the critical thinking lens

Look at the critical thinking lens you constructed and think about the answers to the critical thinking questions in your critical thinking lens. Do you think that the statement "*antacids are bases that neutralize stomach acid*" is a good scientific argument?

Yes ☐   No ☐

Why or why not?

_____

_____

_____

_____

_____

_____

_____

_____

_____

_____

_____

_____

_____

_____

5.5 Now you try

You run into a scientist on the street, and you start talking. He tells you his scientific opinion:

The dish ran away with the spoon.

Evaluate his argument by constructing a critical thinking lens.

Tools for Getting the Facts

Write two questions regarding clarity, precision, accuracy, and detail.

❶ _____

_____

❷ _____

_____

Tools for Evaluating the Facts

Write two questions regarding significance, relevancy, and application.

❶ _____

_____

❷ _____

_____

Tools for Drawing Conclusions (using logic)

Write two questions regarding logical validity, consistency, and flaws.

❶ _____

_____

❷ _____

_____

Tools for Evaluating Conclusions

Write two questions regarding fairness, depth, breadth, and reasonableness.

❶ _____

_____

❷ _____

_____

5.6 Make your own

Using the questions you came up with in section 5.5, construct your own critical thinking lens.

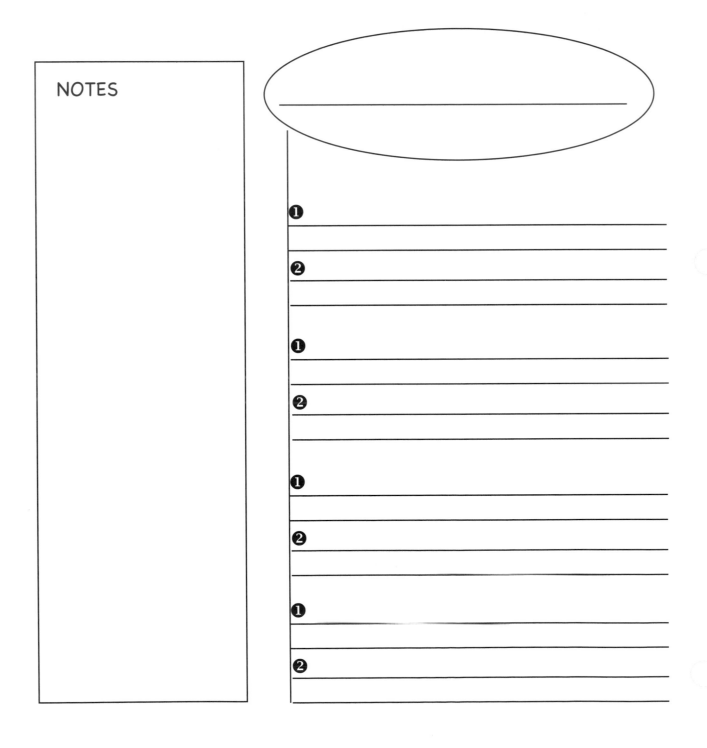

6 Mixtures

Critical Thinking

6.1 Mixtures

6.2 Gathering the tools

6.3 Building the critical thinking lens

6.4 Using the critical thinking lens

6.5 Now you try

6.6 Make your own

6.1 Mixtures

In chapter 6 of Chemistry Level I, you learned about **mixtures.** You also read about the history of **mixtures** and how **mixtures** were discovered. Now you will construct a critical thinking lens to evaluate this scientific claim:

Ice water is a heterogeneous mixture.

[Note: You will need to consider the information you find in your textbook and other resources as "facts" in order to complete this exercise. However, know that the "facts" in your textbook and in other resources are really conclusions that are based on many other facts that have been collected over years of investigation and that have been evaluated by many scientists. As a critical thinker, you are encouraged to examine all "facts" and to evaluate them for yourself, even those facts that have already been evaluated by other scientists.]

6.2 Gathering the tools

First, we need to gather the critical thinking tools. The four types of tools we will be using are as follows:

Tools for Getting the Facts

questions regarding clarity, precision, accuracy, and detail

Tools for Evaluating the Facts

questions regarding significance, relevancy, and application

Tools for Drawing Conclusions (using logic)

questions regarding logical validity, consistency, and flaws.

Tools for Evaluating Conclusions

questions regarding fairness, depth, breadth, and reasonableness

A. Tools for Getting the Facts

questions regarding clarity, precision, accuracy, and detail

Answer the following **Getting the Facts** questions for the statement:

Ice water is a heterogeneous mixture.

1. Q: What is ice water made of?

  A: _____

_____

_____

2. Q: (Write your own question.)

_____

_____

_____

  A: _____

_____

_____

B. Tools for Evaluating the Facts

questions regarding significance, relevancy, and application

Answer the following **Evaluating the Facts** questions for the

statement: *Ice water is a heterogeneous mixture.*

1. Q: Is the fact that ice floats in water relevant to the argument

    that *ice water is a heterogeneous mixture?*

  A: _____

_____

_____

2. Q: (Write your own question.)

_____

_____

_____

  A: _____

_____

C. Tools for Drawing a Conclusion (using Logic)
questions regarding logical validity, consistency, and flaws

In this section, you will learn how to recognize valid arguments and logical fallacies. A logical fallacy is an inaccurate way to formulate an argument. In this chapter, you will be introduced to the logical fallacy called **appeal to pity**.

Logical Fallacy: Appeal to Pity (*Ad Populum*)

Definition: The appeal to pity fallacy is committed when an argument is accepted because not accepting it would cause negative feelings.

Example: I think I should get extra time to do my assignment because my team lost in the playoffs, and I was too upset to work.

To prevent committing the logical fallacy of *appeal to pity,* make sure that the argument is not being accepted because it invokes an emotional response. Evaluate the following arguments for this statement:

Ice water is a heterogeneous mixture.

Mark those using an *appeal to pity.*

☐ Ice water is a heterogeneous mixture because I would feel sad if it wasn't.

☐ Ice water is a heterogeneous mixture because the density of ice is lower than the density of water.

☐ My grandmother always drinks ice water on hot afternoons, and she would be very upset if it wasn't heterogeneous.

☐ A heterogeneous mixture is not the same everywhere. Because ice floats in water, ice water is not the same everywhere, and it is a heterogeneous mixture.

☐ We should not allow more tears in the world, and so we shouldn't say that ice water is heterogeneous.

D. Tools for Evaluating the Conclusion

questions regarding fairness, depth, breadth, and reasonableness

Answer the following **Evaluating the Conclusion** questions for the statement: *Ice water is a heterogeneous mixture.*

1. Q: Is it fair to say that because ice floats in water that *ice water is a heterogeneous mixture?*

A: _____

_____

_____

2. Q: Do you think it is reasonable to conclude that because a heterogeneous mixture is not the same everywhere, *ice water is a heterogeneous mixture?*

A: _____

_____

_____

6.3 Building the critical thinking lens

You have gathered the facts, evaluated the facts, checked the conclusion using logic, and evaluated the conclusion:

Ice water is a heterogeneous mixture.

Next, put all of the facts, evaluations, and logical checks together to construct a critical thinking lens.

Write the statement you are evaluating in the critical thinking lens.

Write the two **Getting the Facts** critical thinking questions.

❶

❷

Write the two **Evaluating the Facts** questions.

❶

❷

Write two **Drawing a Conclusion using Logic** statements that don't commit a logical fallacy.

❶

❷

Write the two **Evaluating the Conclusion** questions.

❶

❷

6.4 Using the critical thinking lens

Look at the critical thinking lens you constructed and think about the answers to the critical thinking questions in your critical thinking lens. Do you think that the statement "*ice water is a heterogeneous mixture*" is a good scientific argument?

☐ Yes   ☐ No

Why or why not?

_____

_____

_____

_____

_____

_____

_____

_____

_____

_____

_____

_____

_____

_____

_____

6.5 Now you try

You run into a scientist on the street, and you start talking. He tells you his scientific opinion:

Grass is green.

Evaluate his argument by constructing a critical thinking lens.

Tools for Getting the Facts

Write two questions regarding clarity, precision, accuracy, and detail.

❶ _____

_____

❷ _____

_____

Tools for Evaluating the Facts

Write two questions regarding significance, relevancy, and application.

❶ _____

_____

❷ _____

_____

Tools for Drawing Conclusions (using logic)

Write two questions regarding logical validity, consistency, and flaws.

❶ _____

_____

❷ _____

_____

Tools for Evaluating Conclusions

Write two questions regarding fairness, depth, breadth, and reasonableness.

❶ _____

_____

❷ _____

_____

6.6 Make your own

Using the questions you came up with in section 6.5, construct your own critical thinking lens.

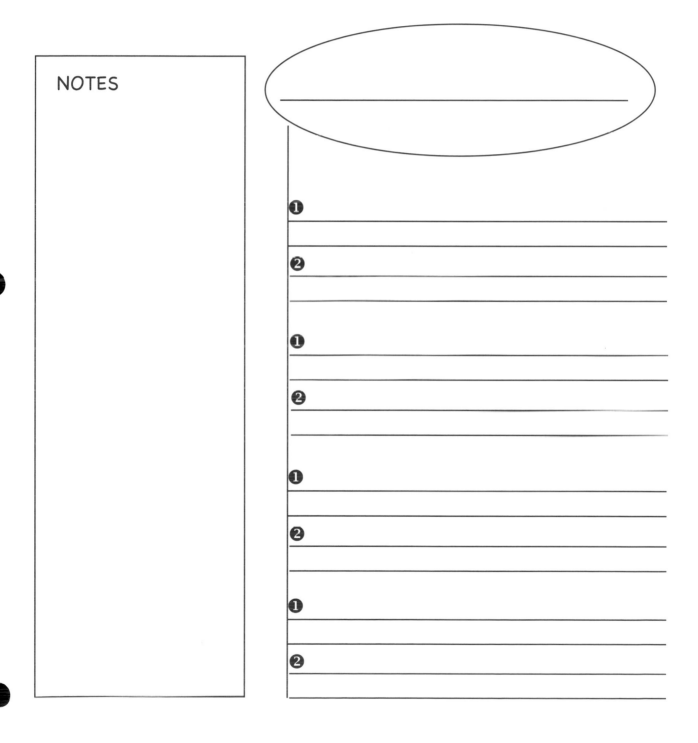

7 Separating Mixtures

Critical Thinking

7.1 Separating mixtures

In chapter 7 of Chemistry Level I, you learned about **separating mixtures.** You also read about the history of **separating mixtures** and how **separation techniques** were discovered.

Now you will construct a critical thinking lens to evaluate this scientific claim:

> *Salt can be separated from water by evaporation.*

[Note: You will need to consider the information you find in your textbook and other resources as "facts" in order to complete this exercise. However, know that the "facts" in your textbook and in other resources are really conclusions that are based on many other facts that have been collected over years of investigation and that have been evaluated by many scientists. As a critical thinker, you are encouraged to examine all "facts" and to evaluate them for yourself, even those facts that have already been evaluated by other scientists.]

7.2 Gathering the tools

First, we need to gather the critical thinking tools. The four types of tools we will be using are as follows:

Tools for Getting the Facts
> *questions regarding clarity, precision, accuracy, and detail*

Tools for Evaluating the Facts
> *questions regarding significance, relevancy, and application*

Tools for Drawing Conclusions (using logic)
> *questions regarding logical validity, consistency, and flaws.*

Tools for Evaluating Conclusions
> *questions regarding fairness, depth, breadth, and reasonableness*

A. Tools for Getting the Facts
questions regarding clarity, precision, accuracy, and detail
Write your own **Getting the Facts** questions and answers for this statement: *Salt can be separated from water by evaporation.*

1. Q: _____

_____

_____

   A: _____

_____

_____

2. Q: _____

_____

_____

   A: _____

_____

_____

B. Tools for Evaluating the Facts
questions regarding significance, relevancy, and application
Write your own **Evaluating the Facts** questions and answers for this statement: *Salt can be separated from water by evaporation.*

1. Q: _____

_____

   A: _____

_____

_____

2. Q: _____

_____

   A: _____

_____

_____

C. Tools for Drawing a Conclusion (using Logic)

questions regarding logical validity, consistency, and flaws

In this section, you will learn how to recognize valid arguments and logical fallacies. A logical fallacy is an inaccurate way to formulate an argument. In this chapter, you will be introduced to the logical fallacy called **appeal to authority**.

Logical Fallacy: Appeal to Authority (*Ad Verecundiam*)

Definition: The appeal to authority happens when a person claims to have authority to say something is true when he or she really does not have this authority.

Example: I think I should get an A on the exam because I have been in school enough years to be a teacher.

To prevent committing the logical fallacy of *appeal to authority,* make sure that the person who is making the argument has the proper authority. Evaluate the following arguments for this statement:

Salt can be separated from water by evaporation.

Mark those using an *appeal to authority.*

☐ I know that salt can be separated from water by evaporation because I did the experiment.

☐ Salt can be separated from water by evaporation because my grandmother told me she heard that somewhere.

☐ Mr. Jones is my neighbor, and he knows a lot of things about cars. He said that you can separate salt from water by evaporation.

☐ The textbook I am using was written by several chemistry professors. They claim salt can be separated from water by evaporation.

☐ My best friend told me that it is true that you can separate salt from water by using evaporation.

D. Tools for Evaluating the Conclusion

questions regarding fairness, depth, breadth, and reasonableness

Write your own **Evaluating the Conclusion** questions and answers for this statement: *Salt can be separated from water by evaporation.*

1. Q: _____

_____

_____

A: _____

_____

_____

2. Q: _____

_____

_____

A: _____

_____

_____

7.3 Building the critical thinking lens

You have gathered the facts, evaluated the facts, checked the conclusion using logic, and evaluated the conclusion:

Salt can be separated from water by evaporation.

Next, put all of the facts, evaluations, and logical checks together to construct a critical thinking lens.

Write the statement you are evaluating in the critical thinking lens.

Write the two **Getting the Facts** critical thinking questions.

Write the two **Evaluating the Facts** questions.

Write two **Drawing a Conclusion using Logic** statements that don't commit a logical fallacy.

Write the two **Evaluating the Conclusion** questions.

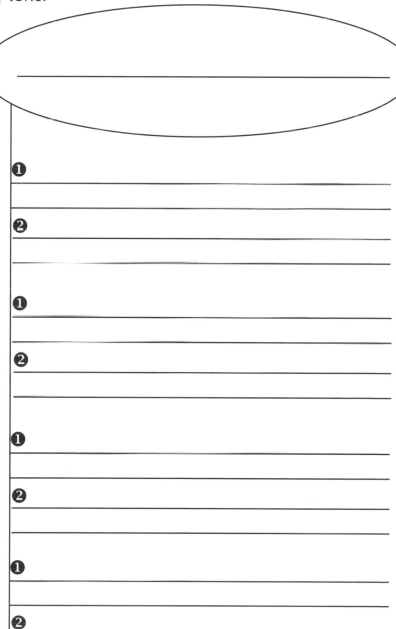

7.4 Using the critical thinking lens

Look at the critical thinking lens you constructed and think about the answers to the critical thinking questions in your critical thinking lens. Do you think that the statement "*salt can be separated from water by evaporation*" is a good scientific argument?

☐ Yes   ☐ No

Why or why not?

_____

_____

_____

_____

_____

_____

_____

_____

_____

_____

_____

_____

_____

_____

7.5 Now you try

You run into a scientist on the street, and you start talking. He tells you his scientific opinion:

The sun always rises in the east.

Evaluate his argument by constructing a critical thinking lens.

Tools for Getting the Facts

Write two questions regarding clarity, precision, accuracy, and detail.

❶ _____

_____

❷ _____

_____

Tools for Evaluating the Facts

Write two questions regarding significance, relevancy, and application.

❶ _____

_____

❷ _____

_____

Tools for Drawing Conclusions (using logic)

Write two questions regarding logical validity, consistency, and flaws.

❶ _____

_____

❷ _____

_____

Tools for Evaluating Conclusions

Write two questions regarding fairness, depth, breadth, and reasonableness.

❶ _____

_____

❷ _____

_____

7.6 Make your own

Using the questions you came up with in section 7.5, construct your own critical thinking lens.

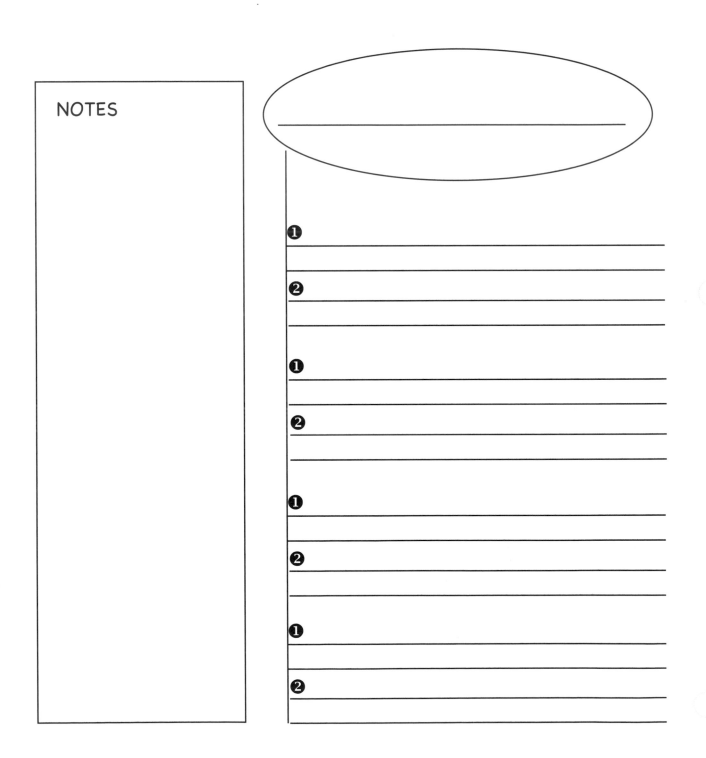

8 Carbohydrates

Critical Thinking

8.1 Carbohydrates

In chapter 8 of Chemistry Level I, you learned about **carbohydrates.** You also read about the history of **carbohydrates** and how **carbohydrates** were discovered. Now you will construct a critical thinking lens to evaluate this scientific claim:

Sucrose is made of a glucose and a fructose.

[Note: You will need to consider the information you find in your textbook and other resources as "facts" in order to complete this exercise. However, know that the "facts" in your textbook and in other resources are really conclusions that are based on many other facts that have been collected over years of investigation and that have been evaluated by many scientists. As a critical thinker, you are encouraged to examine all "facts" and to evaluate them for yourself, even those facts that have already been evaluated by other scientists.]

8.2 Gathering the tools

First, we need to gather the critical thinking tools. The four types of tools we will be using are as follows:

Tools for Getting the Facts
> *questions regarding clarity, precision, accuracy, and detail*

Tools for Evaluating the Facts
> *questions regarding significance, relevancy, and application*

Tools for Drawing Conclusions (using logic)
> *questions regarding logical validity, consistency, and flaws.*

Tools for Evaluating Conclusions
> *questions regarding fairness, depth, breadth, and reasonableness*

A. Tools for Getting the Facts

questions regarding clarity, precision, accuracy, and detail

Write your own **Getting the Facts** questions and answers for this statement: *Sucrose is made of a glucose and a fructose.*

1. Q: _____

_____

_____

A: _____

_____

_____

2. Q: _____

_____

_____

A: _____

_____

_____

B. Tools for Evaluating the Facts

questions regarding significance, relevancy, and application

Write your own **Evaluating the Facts** questions and answers for this statement: *Sucrose is made of a glucose and a fructose.*

1. Q: _____

_____

A: _____

_____

2. Q: _____

_____

A: _____

_____

_____

C. Tools for Drawing a Conclusion (using Logic)
questions of logical validity, consistency, and flaws

In this section, you will learn how to recognize valid arguments and logical fallacies. A logical fallacy is an inaccurate way to formulate an argument. In this chapter, you will be introduced to the logical fallacy called **red herring**.

Logical Fallacy: Red Herring

Definition: A red herring fallacy occurs when an irrelevant topic is introduced to divert attention from the original argument.

Example: I don't think I should have to turn in my homework because we just moved to a new house.

To prevent committing the logical fallacy of *red herring,* make sure that the argument does not include irrelevant information.
Evaluate the following arguments for this statement:

Sucrose is made of a glucose and a fructose.
Mark those using a *red herring.*

☐ I know that sucrose is made of glucose and fructose because carbohydrates contain water.

☐ Sucrose is a sugar with two monosaccharides called glucose and fructose.

☐ Sucrose is made of glucose and fructose because sucrose tastes great on strawberries.

☐ Glucose and fructose are linked together by a chemical bond to make sucrose - an oligosaccharide.

☐ Sucrose is used to make pies and cakes, so it is made of glucose and fructose.

D. Tools for Evaluating the Conclusion

questions of fairness, depth, breadth, and reasonableness

Write your own **Evaluating the Conclusion** questions and answers for this statement: *Sucrose is made of a glucose and a fructose.*

1. Q: _____

_____

_____

A: _____

_____

_____

2. Q: _____

_____

_____

A: _____

_____

_____

8.3 Building the critical thinking lens

You have gathered the facts, evaluated the facts, checked the conclusion using logic, and evaluated the conclusion:

Sucrose is made of glucose and fructose.

Next, put all of the facts, evaluations, and logical checks together to construct a critical thinking lens.

Write the statement you are evaluating in the critical thinking lens.

Write the two **Getting the Facts** critical thinking questions.

❶

❷

Write the two **Evaluating the Facts** questions.

❶

❷

Write two **Drawing a Conclusion using Logic** statements that don't commit a logical fallacy.

❶

❷

Write the two **Evaluating the Conclusion** questions.

❶

❷

8.4 Using the critical thinking lens

Look at the critical thinking lens you constructed and think about the answers to the critical thinking questions in your critical thinking lens. Do you think that the statement "*sucrose is made of glucose and fructose*" is a good scientific argument?

☐ Yes   ☐ No

Why or why not?

_____

_____

_____

_____

_____

_____

_____

_____

_____

_____

_____

_____

_____

8.5 Now you try

You run into a scientist on the street, and you start talking. He tells you his scientific opinion:

Snowflakes are all unique.

Evaluate his argument by constructing a critical thinking lens.

Tools for Getting the Facts

Write two questions regarding clarity, precision, accuracy, and detail.

❶ _____

_____

❷ _____

_____

Tools for Evaluating the Facts

Write two questions regarding significance, relevancy, and application.

❶ _____

_____

❷ _____

_____

Tools for Drawing Conclusions (using logic)

Write two questions regarding logical validity, consistency, and flaws.

❶ _____

_____

❷ _____

_____

Tools for Evaluating Conclusions

Write two questions regarding fairness, depth, breadth, and reasonableness.

❶ _____

_____

❷ _____

_____

8.6 Make your own

Using the questions you came up with in section 8.5, construct your own critical thinking lens.

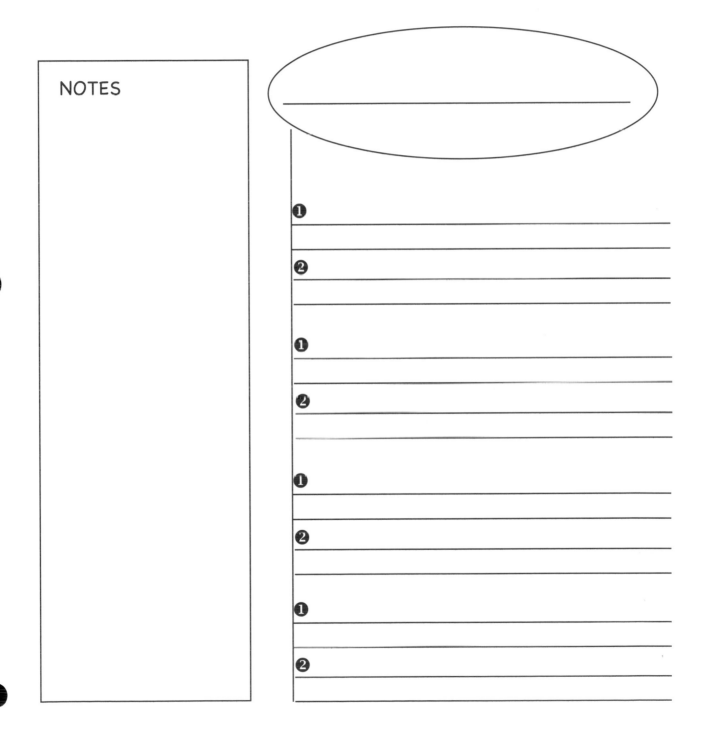

9 Polymers

Critical Thinking

9.1 Polymers

In chapter 9 of Chemistry Level I, you learned about **polymers.** You also read about the history of **polymers** and how **polymers** were discovered. Now you will construct a critical thinking lens to evaluate this scientific claim:

Plastics are made of polymers.

[Note: You will need to consider the information you find in your textbook and other resources as "facts" in order to complete this exercise. However, know that the "facts" in your textbook and in other resources are really conclusions that are based on many other facts that have been collected over years of investigation and that have been evaluated by many scientists. As a critical thinker, you are encouraged to examine all "facts" and to evaluate them for yourself, even those facts that have already been evaluated by other scientists.]

9.2 Gathering the tools

FIrst, we need to gather the critical thinking tools. The four types of tools we will be using are as follows:

Tools for Getting the Facts
> *questions regarding clarity, precision, accuracy, and detail*

Tools for Evaluating the Facts
> *questions regarding significance, relevancy, and application*

Tools for Drawing Conclusions (using logic)
> *questions regarding logical validity, consistency, and flaws.*

Tools for Evaluating Conclusions
> *questions regarding fairness, depth, breadth, and reasonableness*

A. Tools for Getting the Facts

questions regarding clarity, precision, accuracy, and detail

Write your own **Getting the Facts** questions and answers for this statement: *Plastics are made of polymers.*

1. Q: _____

_____

_____

A: _____

_____

_____

2. Q: _____

_____

_____

A: _____

_____

_____

B. Tools for Evaluating the Facts

questions regarding significance, relevancy, and application

Write your own **Evaluating the Facts** questions and answers for this statement: *Plastics are made of polymers.*

1. Q: _____

_____

A: _____

_____

2. Q: _____

_____

A: _____

_____

C. Tools for Drawing a Conclusion (using Logic)
questions of logical validity, consistency, and flaws

In this section, you will learn how to recognize valid arguments and logical fallacies. A logical fallacy is an inaccurate way to formulate an argument. In this chapter, you will be introduced to the logical fallacy called **bandwagon**.

Logical Fallacy: Bandwagon

Definition: Bandwagon is a fallacy that occurs when a person faces rejection from a group if he or she doesn't agree.

Example: Jane, you may believe that the sky is blue, but that's not true because we don't believe that in these parts.

To prevent committing the logical fallacy of *bandwagon,* make sure that the argument does not try to exclude someone from the group. Try writing your own arguments for this statement:

Plastics are made of polymers.

(Write an argument using the bandwagon fallacy.)

_____

_____

_____

(Write an argument using the bandwagon fallacy.)

_____

_____

(Write an argument without committing the bandwagon fallacy.)

_____

_____

_____

(Write an argument using the bandwagon fallacy.)

_____

_____

_____

(Write an argument without committing the bandwagon fallacy.)

_____

_____

_____

D. Tools for Evaluating the Conclusion
questions regarding fairness, depth, breadth, and reasonableness

Write your own **Evaluating the Conclusion** questions and answers for this statement: _Plastics are made of polymers._

1. Q: _____

_____

_____

   A: _____

_____

_____

2. Q: _____

_____

_____

   A: _____

_____

9.3 Building the critical thinking lens

You have gathered the facts, evaluated the facts, checked the conclusion using logic, and evaluated the conclusion:

Plastics are made of polymers.

Next, put all of the facts, evaluations, and logical checks together to construct a critical thinking lens.

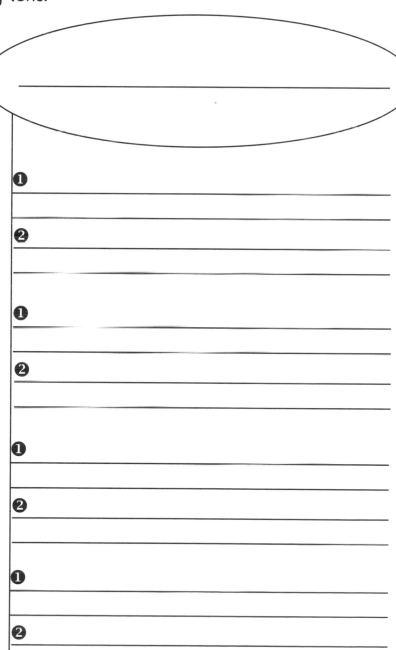

Write the statement you are evaluating in the critical thinking lens.

Write the two **Getting the Facts** critical thinking questions.

Write the two **Evaluating the Facts** questions.

Write two **Drawing a Conclusion using Logic** statements that don't commit a logical fallacy.

Write the two **Evaluating the Conclusion** questions.

9.4 Using the critical thinking lens

Look at the critical thinking lens you constructed and think about the answers to the critical thinking questions in your critical thinking lens. Do you think that the statement "*plastics are made of polymers*" is a good scientific argument?

Yes ☐   No ☐

Why or why not?

_____

_____

_____

_____

_____

_____

_____

_____

_____

_____

_____

_____

_____

9.5 Now you try

You run into a scientist on the street, and you start talking. He tells you his scientific opinion:

(Write a statement you have heard someone say.)

_____

Evaluate his argument by constructing a critical thinking lens.

Tools for Getting the Facts

Write two questions regarding clarity, precision, accuracy, and detail.

❶ _____

_____

❷ _____

_____

Tools for Evaluating the Facts

Write two questions regarding significance, relevancy, and application.

❶ _____

_____

❷ _____

_____

Tools for Drawing Conclusions (using logic)

Write two questions regarding logical validity, consistency, and flaws.

❶ _____

_____

❷ _____

_____

Tools for Evaluating Conclusions

Write two questions regarding fairness, depth, breadth, and reasonableness.

❶ _____

❷ _____

_____

9.6 Make your own

Using the questions you came up with in section 9.5, construct your own critical thinking lens.

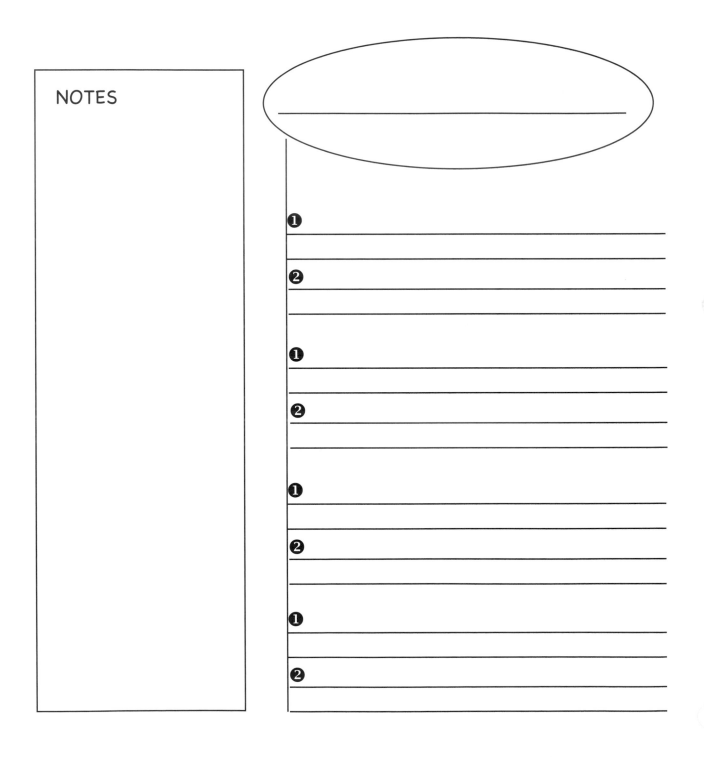

NOTES

❶

❷

❶

❷

❶

❷

❶

❷

10 Biological Polymers

Critical Thinking

10.1 Biological polymers

In chapter 10 of Chemistry Level I, you learned about **biological polymers.** You also read about the history of **biological polymers** and how **polymers** were discovered. Now you will construct a critical thinking lens to evaluate this scientific claim:

Proteins are made of amino acids.

[Note: You will need to consider the information you find in your textbook and other resources as "facts" in order to complete this exercise. However, know that the "facts" in your textbook and in other resources are really conclusions that are based on many other facts that have been collected over years of investigation and that have been evaluated by many scientists. As a critical thinker, you are encouraged to examine all "facts" and to evaluate them for yourself, even those facts that have already been evaluated by other scientists.]

10.2 Gathering the tools

First, we need to gather the critical thinking tools. The four types of tools we will be using are as follows:

Tools for Getting the Facts
   questions regarding clarity, precision, accuracy, and detail
Tools for Evaluating the Facts
   questions regarding significance, relevancy, and application
Tools for Drawing Conclusions (using logic)
   questions regarding logical validity, consistency, and flaws.
Tools for Evaluating Conclusions
   questions regarding fairness, depth, breadth, and reasonableness

A. Tools for Getting the Facts

questions regarding clarity, precision, accuracy, and detail

Write your own **Getting the Facts** questions and answers for this statement: *Proteins are made of amino acids.*

1. Q: _____

_____

_____

  A: _____

_____

_____

2. Q: _____

_____

_____

  A: _____

_____

_____

B. Tools for Evaluating the Facts

questions regarding significance, relevancy, and application

Write your own **Evaluating the Facts** questions and answers for this statement: *Proteins are made of amino acids.*

1. Q: _____

_____

_____

  A: _____

_____

_____

2. Q: _____

_____

_____

  A: _____

_____

_____

C. Tools for Drawing a Conclusion (using Logic)

questions of logical validity, consistency, and flaws

In this section, you will learn how to recognize valid arguments and logical fallacies. A logical fallacy is an inaccurate way to formulate an argument. In this chapter, you will be introduced to the logical fallacy called **appeal to belief**.

Logical Fallacy: Appeal to Belief

Definition: Appeal to belief is a fallacy that occurs when something is considered true because most people believe it is.

Example: Most people believe that it is better to use hot water to wash clothes. That must be true.

To prevent committing the logical fallacy of *appeal to belief,* make sure that the argument is not making a claim based on what most people believe. Try writing your own arguments for this statement:

Proteins are made of amino acids.

(Write an argument using the appeal to belief fallacy.)

_____

_____

_____

(Write an argument using the appeal to belief fallacy.)

_____

_____

_____

(Write an argument without committing the appeal to belief fallacy.)

_____

_____

_____

(Write an argument using the appeal to belief fallacy.)

_____

_____

_____

(Write an argument without committing the appeal to belief fallacy.)

_____

_____

_____

D. Tools for Evaluating the Conclusion

questions regarding fairness, depth, breadth, and reasonableness

Write your own **Evaluating the Conclusion** questions and answers for this statement: _Proteins are made of amino acids._

1. Q: _____

_____

A: _____

_____

2. Q: _____

_____

A: _____

_____

10.3 Building the critical thinking lens

You have gathered the facts, evaluated the facts, checked the conclusion using logic, and evaluated the conclusion:

Proteins are made of amino acids.

Next, put all of the facts, evaluations, and logical checks together to construct a critical thinking lens.

Write the statement you are evaluating in the critical thinking lens.

Write the two **Getting the Facts** critical thinking questions.

❶

❷

Write the two **Evaluating the Facts** questions.

❶

❷

Write two **Drawing a Conclusion using Logic** statements that don't commit a logical fallacy.

❶

❷

Write the two **Evaluating the Conclusion** questions.

❶

❷

10.4 Using your critical thinking lens

Look at the critical thinking lens you constructed and think about the answers to the critical thinking questions in your critical thinking lens. Do you think that the statement *"proteins are made of amino acids"* is a good scientific argument?

Yes ☐   No ☐

Why or why not?

_____

_____

_____

_____

_____

_____

_____

_____

_____

_____

_____

_____

_____

_____

10.5 Now you try

You run into a scientist on the street, and you start talking. He tells you his scientific opinion:

(Write a statement you have heard someone say.)

Evaluate his argument by constructing a critical thinking lens.

Tools for Getting the Facts

Write two questions regarding clarity, precision, accuracy, and detail.

❶ _____

_____

❷ _____

_____

Tools for Evaluating the Facts

Write two questions regarding significance, relevancy, and application.

❶ _____

_____

❷ _____

_____

Tools for Drawing Conclusions (using logic)

Write two questions regarding logical validity, consistency, and flaws.

❶ _____

_____

❷ _____

_____

Tools for Evaluating Conclusions

Write two questions regarding fairness, depth, breadth, and reasonableness.

❶ _____

_____

❷ _____

_____

10.6 Make your own

Using the questions you came up with in section 10.5, construct your own critical thinking lens.

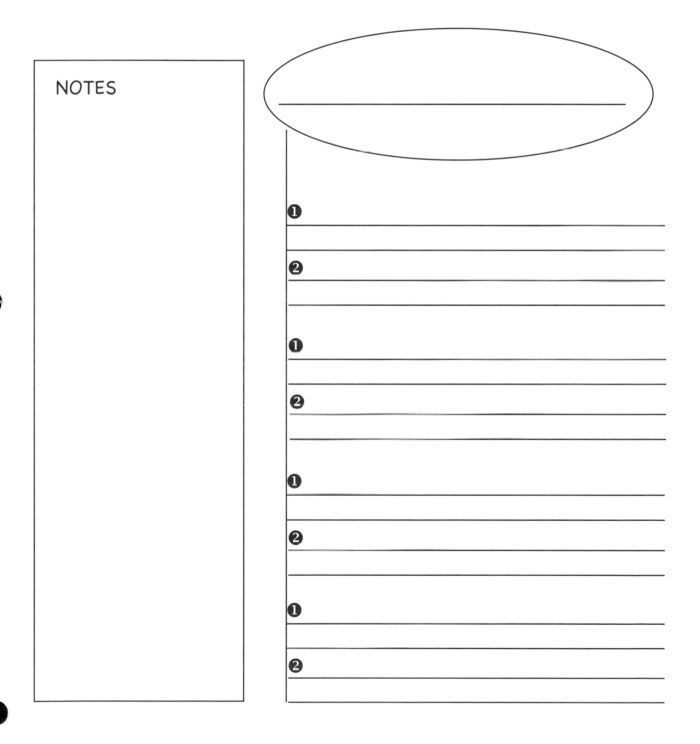

Do you need more room to write your answers? Use these pages.

NOTES

❶

❷

❶

❷

❶

❷

❶

❷

NOTES

❶

❷

❶

❷

❶

❷

❶

❷